Fahad Alghamdi

Vernacular Landscape

Fahad Alghamdi

Vernacular Landscape

A Story from Saudi Arabia

AV Akademikerverlag

Impressum / Imprint

Bibliografische Information der Deutschen Nationalbibliothek: Die Deutsche Nationalbibliothek verzeichnet diese Publikation in der Deutschen Nationalbibliografie; detaillierte bibliografische Daten sind im Internet über http://dnb.d-nb.de abrufbar.
Alle in diesem Buch genannten Marken und Produktnamen unterliegen warenzeichen-, marken- oder patentrechtlichem Schutz bzw. sind Warenzeichen oder eingetragene Warenzeichen der jeweiligen Inhaber. Die Wiedergabe von Marken, Produktnamen, Gebrauchsnamen, Handelsnamen, Warenbezeichnungen u.s.w. in diesem Werk berechtigt auch ohne besondere Kennzeichnung nicht zu der Annahme, dass solche Namen im Sinne der Warenzeichen- und Markenschutzgesetzgebung als frei zu betrachten wären und daher von jedermann benutzt werden dürften.

Bibliographic information published by the Deutsche Nationalbibliothek: The Deutsche Nationalbibliothek lists this publication in the Deutsche Nationalbibliografie; detailed bibliographic data are available in the Internet at http://dnb.d-nb.de.
Any brand names and product names mentioned in this book are subject to trademark, brand or patent protection and are trademarks or registered trademarks of their respective holders. The use of brand names, product names, common names, trade names, product descriptions etc. even without a particular marking in this works is in no way to be construed to mean that such names may be regarded as unrestricted in respect of trademark and brand protection legislation and could thus be used by anyone.

Coverbild / Cover image: www.ingimage.com

Verlag / Publisher:
AV Akademikerverlag
ist ein Imprint der / is a trademark of
OmniScriptum GmbH & Co. KG
Heinrich-Böcking-Str. 6-8, 66121 Saarbrücken, Deutschland / Germany
Email: info@akademikerverlag.de

Herstellung: siehe letzte Seite /
Printed at: see last page
ISBN: 978-3-639-72142-3

Copyright © 2014 OmniScriptum GmbH & Co. KG
Alle Rechte vorbehalten. / All rights reserved. Saarbrücken 2014

ABSTRACT

A vernacular landscape is one that has naturally developed through the everyday use of a certain space. It reveals the relationship between humans and nature in a particular time in history and can be found in the interaction between human cultures, the built environment, and the natural environment. Over time, ordinary landscapes have evolved, in part because of development by human beings. These man-made changes resulted in new forms and patterns in the vernacular landscape.

This study explores the basic concept of the vernacular landscape in the context of the relationship between landscape form and culture across time. The discussion of different meanings of the word "vernacular" and "vernacular landscape" are provided, in addition to explanations about why the concept is important to understand. The study should advance the debate about landscape without landscape architects and how landscape architects read a landscape in historical context.

This study will provide a view of the southwestern region of Saudi Arabia where the vernacular landscape has dramatically changed in the last few decades. The paper discusses the influence of different factors on the transformation process and presents selected Islamic principles and traditional norms that preserve this particular vernacular landscape over time.

This paper also includes one case study about Hezna village as prototype for the transformation of the vernacular landscape in the region. Oral interviews and fields investigations have been conducted in order to clearly describe the transformation process and the history of degradation. In addition, visual assessments were done using photography and aerial images to trace the physical changes in the vernacular landscape in the southwestern region of Saudi Arabia.

ACKNOWLEDGMENTS

The author wishes to express sincere appreciation to Professors Roman Lenz and Dr. Ahmad Al-Gilani for their assistance, advice and patience. In addition, special thanks to Dr. Ahmad Qushas for his valuable input and discussion. Thanks also to my colleagues Abulgader Alamri and Saed Aref for their support and encouragements.

TABLE OF CONTENTS

LIST OF FIGURES

Chapter 1

INTRODUCTION

1.1 Statement of the Problem

The transformation of vernacular landscape is no longer associated with new agriculture technologies, which can create new patterns of landscape. Social, economic, and political changes can influence the transformation of the vernacular landscape. In the mid-1960s, Saudi Arabia experienced a significant shift in the history of the vernacular landscape. The transformation of the tribal land management system, combined with the economic and population growth of the country, led to dynamic change in the traditional structure and the physical appearance of the vernacular landscape. As a result, in just one generation, agricultural practices were lost, dwelling areas expanded into important agriculture land, and traditional land management became a relic.

1.2 Research Questions

This study attempts to critically examine the fundamental principles of the vernacular landscape in order to evaluate the transformation process in Saudi Arabia. It attempts to clarify the main factors behind the transformation in order to evaluate the influence of different factors in the overall image of the vernacular landscape. To a large extent, the causes and effects of different factors can be applied in one specific case study from the region. The case study includes evaluation of the existing conditions and predictions for the vernacular landscape.

1.3 Methodical Approach

The hypothesis of this study is that traditional societies have developed a set of strong norms and rules that tend to preserve the vernacular landscape for centuries. The transformation started when administrative authorities took over the responsibilities of the land management and natural resources. The verification of this assumption was approached through objective, systematic analysis and field investigation at the study location. Aspects considered include the location's functional qualities, historical development, sociological claims and spatial qualities. In addition, oral interviews replaced the lack of written material regarding historical degradation, while photography highlights the physical evidence of change in the vernacular landscape.

1.4 Scope

The analytical investigation of the concept of vernacular landscape can be examined within the scope of landscape planning and landscape polices. In the case of Saudi Arabia, the transformation in the vernacular landscape and traditional land management system possesses similar characteristics especially in the southwestern part of the country. In addition, a case study about a village which can be a prototype for the transformation of the vernacular landscape in the region. In this scale, more in-depth research can be conducted in order to clarify the transformation process.

Chapter 2

LITERATURE REVIEW

2.1 Landscape

According to Jackson (1984), Landscape is a ubiquitous word appearing in English and other European languages, with origins in Anglo-German language dating back to c. 500AD in Europe. The words *landskipe,* or *landscaef,* and the notions implied, were taken to Britain by Anglo-Saxon settlers.[1] The words translated to a clearing in the forest with animals, huts, fields, and fences. It was essentially a peasant landscape carved out of the original forest, or *weald* (wilderness), with interconnections to patterns of occupation and associated customs and ways of doing things. Landscape, from its beginnings, has meant a man-made artifact with associated cultural process values. Its morphology results from the interplay between cultural values, customs, and land-use practices. Recently explored by Wylie (2007), it is what he calls "an active scene of practice."[2]

It also has, as Jackson (1984) indicates, the equivalent meaning in Latin-based languages, with antecedents like Germanic and other languages harking back to the Indo-European idiom. It is derived from the Latin *pagus,* meaning a defined rural district. Jackson notes that this gives the French words *pays* and *paysage,* but that there are other French words for landscape including *champagne,* deriving from champagne, meaning a countryside of fields. The English equivalent was once "champion."[3]

Briefly stated, the principle requires landscape to be a fusion of two major perspectives: functional and moral-aesthetic. Originally, the term "landscape" referred primarily, to the workday world, to an estate or a domain. From the

3

sixteenth century on—particularly in the Netherlands and England—landscape acquired increasingly more of an aesthetic meaning; it became a genre of art.[4]

From this piece of information, Jackson argued that:

> We can learn two things. First, that our Dark Age forebears possessed skills which we probably did not credit them with, and second, that the word *scape* could also indicate something like an organization or a system. And why not? If housescape meant the organization of the personal of a house, if township eventually came to mean an administrative unit, then landscape could well have meant something like an organization, a system of rural farm spaces. At all events it is clear that a thousand years ago the word had nothing to do with scenery or the depiction of scenery.[5]

In the same context, Meining (1979) also stated that " it Brings with it not only a new kind of but a new concept of landscape, a new kind of architecture and a new kind of urban design. Both in the country and the city it will, we believe, produce what we so badly need: something can be described as esthetic planning, planning for delight and stimulation of the sense as well as of the spirit."[6]

Nevertheless, Jackson (1984) further argued that, the idea of landscape as man-made spaces on a land reveal a concept, and it does not provide us with definition of landscape. Therefore, landscape is functioning and evolving, not according to natural process but to serve a human needs. He stated, "The collective character of the landscape is one thing that all generations and all points of view agreed upon. A landscape is thus a space deliberately created to speed up or slow down the process of nature".[7]

From landscape architects' point of view, the definition of landscape gradually developed to be more related to people and their interaction with nature. According to Halprin (1995), landscape is "nature converted into an opera invested with human passion, so it becomes functionally and emotionally related to the needs and uses to which we put nature."[8]

The definition of landscape used by the European Landscape Convention confirms the necessity of human factors by defining landscape as "an area, as perceived by people, whose character is the result of the action and interaction of natural and/or human factors."[9]

2.2 Vernacular

According to Jackson (1984), The word derives from the Latin *verna*, meaning a slave born in the house of his or her master, and by extension, in classical times, it meant a native, one whose existence was confined to a village or estate and who was devoted to routine work. A vernacular culture would imply a way of life ruled by tradition and custom, entirely remote from the larger world of politics and law; a way of life where identity derived not from permanent possession of land, but from membership in a group or super family.[10]

The dictionaries meaning of vernacular is ordinary and every day. It is often associated with the language, culture, and history. These are some definitions of the word "vernacular" from different dictionaries and resources. One definition describes the word vernacular as "slang: a characteristic language of a particular group" or "the everyday speech of the people (as distinguished from literary language)." [11]

In terms of architecture, vernacular architecture has been defined as "The traditional architecture of a region, frequently developed in response to the climate, land conditions, or culture of a region." [12] In the same context, vernacular architecture could be for one specific building described as "A building built without being designed by an architect or engineer or someone with similar formal training, often based on traditional or regional forms." [13]

Some definitions expand the meaning of the word to include "everyday expression of cultural groups, from language to architecture." [14] As landscape architects, we can modify this, which is to say that vernacular is the "every day expression of cultural groups, from language to landscape."

2.3 Vernacular Landscape

Figure 1 Sacareni village, Romania

When the topic of the vernacular landscape is discussed, the writings of J. B. Jackson should be mentioned. Jackson defined the concept for the field of landscape architecture. In his 1984 work, *Discovering the Vernacular Landscape*, he discusses in-depth the idea of vernacular landscape as historical layers representing the story of the landscape's transformation. He wrote that, "Only very rarely is there a glimpse of the history of the landscape itself, how it was formed, how it has changed, and who it was, who changed it, and even more rarely does landscape research produce any speculation about the nature of the landscape." [15] Jackson (1984) continues:

> Change in itself is not out of the ordinary; every cultural landscape has evolved, sometimes violently, more often slowly, over the centuries. What differs here is that we are able to watch the transformation as it takes place; able to record it and even to understand some of its signs. [16]

Therefore, the vernacular landscape can be considered an integration of man-made landscape, which consists of the built, agricultural, and natural landscape. Agriculture has historically been the main factor influencing the character of the vernacular landscape. Many of the fields, their boundaries, and the villages near them reflect the pattern of agricultural activity in the past. The vernacular landscape can thus be said "to express the relationship between a human group and the land it occupies." [17]

Another resource is the collection edited by Paul Groth and Todd W. Bresi (1994), in which a cultural geographer's view of landscapes is described as a "...focus on the history of how people have used everyday space – buildings, rooms, streets, fields, or yards – to establish their identity, articulate their social relations, and derive cultural meaning." [18]

But in a more traditional context, the relationship between human and nature is more equal, as human habitation relies more on natural systems and less on technological achievement. In either case, one of the most important characteristics of a vernacular landscape is its ability to express the needs of the human group that occupies it. Since it must sustain the group both physically and psychologically, its design mirrors their social, aesthetic, political, economic and spiritual values. [19]

According to Jackson (1984), it is often the legal aspects of the landscape that give us the clearest insight; especially into the relationship between the peasant or villager and the piece of land he works. [20]

For the vernacular landscape's construction and evolution, people and nature are the two main players. People are certainly dominant, and nature is brought into settlement for many purposes.

2.4 Reading Landscape and Landscape Archive

According to Lewis (1979), the basic principle is that "all human landscape has cultural meaning," no matter how ordinary that landscape may be. It follows, as Mae Thielgaard Watts has remarked, that we can "read the landscape" as we might read a book. Grady Clay said it well: "There are no secrets in the landscape." All of our culture's "warts and blemishes" are there, but so are our glories. Above all, our ordinary day-to-day qualities are exhibited for anybody who wants to find them and knows how to look for them. Lewis (1979), however, argues that reading a landscape is not as easy as reading a book, and the reason is that "ordinary landscape seems messy and disorganized, like a book with pages missing, torn, and smudged; a book whose copy has been edited and re-edited by people with illegible handwriting. Like books, landscapes can be read, but unlike books, they were not meant to be read." [21]

In discussing the origin of his book, Hoskin (1973) remembered how he looked at the landscape during his childhood when he wrote, "Even then, I felt that everything I was looking at was saying something to me if I could recognize the language. It was a landscape written in a kind of code." [22]

2.5 The Basics of Reading Landscape

In his work, Lewis (1979) argued that, very few academic disciplines teach their students how to read landscape, or encourage them to try. Traditional geomorphology and traditional plant ecology were two happy exceptions: these disciplines insisted that their practitioners use their eyes and think about what they saw, and it is no accident that some of America's most accomplished landscape readers, such as J. Hoover Mackin, Pierre Dansereau, and Mae Thielgaard Watts, come from those fileds.[23]

This is not always the case, however. People like J. B. Jackson, the founder and editor of Landscape Magazine, one of the talented landscape readers, is not attached to any discipline; although his work has influence and changed the way many look at and read American landscape.

Reading landscape is a public or "human art" as D. W. Meing (1979) called it. Meing said, "Reading landscape is a human art, unrestricted to any profession, unbounded by any field, unlimited in its challenges and pleasures." [24] Hoskin also agreed that visible landscape itself "offers us enough stimulus and pleasure without uncertainty about what may lie underneath."

Reading vernacular landscape requires basic knowledge in landscape history and locality. A person should be able to "look at every feature with exact knowledge, able to give a name to it and know how it got there, and not just gaze uncomprehendingly at it as a beautiful but silent view." [25]

However, reading landscape is one thing, but the ability to understand its signs and meaning is often something else. For instance, a non-native reader might be able to read any set of German words, but interpreting a passage from a German text is often more difficult. Thus, other than the ability to read landscape, one should have a degree of belonging to a place in order to be able to understand the foundational workings of daily life in that place.

For Jackson, he can read a stone walls around the watershed in the southern region of Saudi Arabia as a political landscape element or boundaries, which is true; but he should live there for a while to understand this is not only shows where the property end; but it shows where the bloodshed start. At any rates, one should first learn how to read and understand landscape in order to be able to deal with it or to decide what could be the best intervention, if needed, to sustain the natural process.

2.6 "Logics" or "Rrules" for Rreading Landscape.....

There is no doubt that ordinary people and professionals need some help and guidance to read landscape. Unless " one is lucky enough to have studied with a plant ecologist like Dansereau, a geomorphologist like Mackin, a folklorist like Glassie, or simply a Renaissance man like Jackson, one is likely need a guidance."[26]

After a significant amount of time spent learning and teaching cultural landscape, Lewis (1979) developed a set of rules or axioms to aid in reading and understanding a landscape. It is worth discussing these "rules" and their corollaries, even if the terms "rules" or "axioms" are not applied. He has stated, " I may be wrong in using the word "axiom": what seems self-evident now was not obvious to me a few years ago. But call them what you will: they are nevertheless essential ideas that underline the reading of America's cultural landscape." [27]

Lewis's "axioms" for reading landscape can be summarized as follows: [28]

2.6.1 Landscape as clue to culture.
Lewis meant that "the culture of any nation is unintentionally reflected in its ordinary vernacular landscape." This fact includes some corollaries:

- **The corollary of cultural change**: If there is a major change in the look of the vernacular landscape, then there is a major change occurring in the national cultural at the same time.

- **The regional corollary:** If in one part of the country, a city looks substantially different than in another, there is a good chance that the cultures of the two places are different as well.

- **The corollary of convergence:** In contrast, if the look of two places is similar, then one can surmise that the cultures are converging.

- **The corollary of diffusion:** The look of the landscape is often changed by imitation. People imitate what they like somewhere else. Timing and location of imitative changes tell a good deal about social diffusion and how cultural ideas spread and change.

- **The corollary of taste:** Different cultures possess different tastes in cultural landscape; to understand the root of taste is to understand much of the culture itself.

2.6.2 Cultural unity and landscape equality: Every element in the human landscape reflects culture in some way. Therefore, every item has an equal importance as a clue to culture. Thus, Lewis argues that "a McDonald's hamburger stand is just as important a cultural symbol or (clue) as the Empire State Building." [29]

2.6.3 Common things: The common landscape by its nature is somehow hard to study by conventional academic means. This is because it is usually neglected and combined with snobbery.

- **The corollary of nonacademic literature:** For the study of vernacular landscape, there is a treasury of information one can get from new journalists, trade journals, advertisements, commercial products, and from promotional travel literature.

2.6.4 The historic element: A large part of the common landscape was built by people in the past, whose taste, habits, technology, wealth, and ambitions were different from ours. Thus, while we live among items from the past, we must try to understand the people who built them in their cultural context, not ours.

- **The corollary of historic lumpiness:** Most of the major change in the cultural landscape does not occur gradually. Instead, it happens in sudden historic leaps, such as through war, depression, and major innovation. These historical time shifts are the key factor to understanding why a landscape changes quickly and drastically.

- **The mechanical, technological corollary:** To understand the cultural significance of a landscape, it is essential to know something in particular about the mechanics of technology and communication that made the landscape elements possible.

2.6.5 The geographic or ecological element: Because of the significant role of culture in landscape, landscape makes little sense if studied out of its geographic and ecological context.

2.6.6 Environmental control: Contemporary technology is so powerful that we can build virtually anything wherever we like, and in so doing, run the risk of ignoring climate, landform, soil, and so on. However, most cultural landscapes, especially those formed in the past, are intimately related to the physical environment. Thus, the reading of cultural landscape also requires basic knowledge in physical landscape.

2.6.7 Landscape obscurity: "Most of the objects in a landscape, although they convey all kind of messages, do not convey those messages in any obvious way." Landscape does not speak to us very clearly. Or, we should at least know what kind of question to ask of it. Reading landscape will not answer every question related to culture, but it should be a clue, an indicator, or evidence in some cases, to help understand how people lived in a place at a particular time.

In spite the fact that the author here was talking about American cultural landscape, these rules can be applied to any cultural landscape. However, these rules are not exhaustive; some cultures are rich enough to develop their own rules and logic.

2.7 Landscape as Biography

Often, there is a logic or rational basis for everything formed in the vernacular landscape; if one could not find it that basis, it does not mean one does not exist. For instance, we have assumptions and theories about how ancient Egyptians lived and developed their own landscape, or how and why the Great Wall of China was built in a certain way, or why Stonehenge was located where it was.

The question is why we need to know the people who were created a given landscape. Marwyn Samuels (1979) tried to answer this question. He wrote, "If we cannot know who they were, we can at least make known who they were not. Perhaps, in those cases, that is all what we can ever hope to accomplish." [30]

Some vernacular landscapes have been occupied by many cultures over the time without leaving any human signs. This certainly makes it more complicated to understand or even renders the landscape unreadable due to the lack of usable information. However, this should not erode the attempt to understand vernacular landscape.

In this sense, Samuels (1979) argued that the biography of landscape is limited in its applicability; however, it is limited by the lack of concrete data in all cases, and by the willingness to search diligently enough. He stated, "This should hardly bother those of us who profess loyalty to science or to the desire to explain the landscape of man. What else, other than the data and a willingness to search out the data, should limit understanding." [31]

2.8 Landscape Without the Landscape Architect

Landscape can yield different interpretations in different places. The definition and perception of a landscape has been formed based on the relationship between humans and nature, and that relationship changes over time. The vernacular landscape is the landscape that has been shaped by the interrelation of human need and by the availability of natural resources, with little or no regard for the organizational quality. Therefore, this landscape exists without landscape architect. The question here is what landscape architecture has done to that historically organic human-environment interaction. Is a modern design inherently vernacular, or it has been designed to be vernacular landscape?

Returning to the concept of vernacular, Jackson's theory consists of Landscapes One, Two, and Three, which form an evolution of thinking and interpretation of the vernacular. Landscape One is essentially the medieval landscape; it is formed by the intermingling of spaces and forms without particular organizational qualities. Landscape Two, or the Renaissance, consists of single-purposed designed landscapes and "... sets great score on visibility; that is why we have that seventeenth-century definition of landscape as 'a vista or view of scenery of the land' – landscape as a work of art, as a kind of super garden." [32]

Landscape Three is the ephemeral *place* we discuss and strive for, requiring not just space, but the connection to humanity. According to Jackson, landscape three can be described as:

> Far greater degree, we derive our identity from our relationship with other people, and when we talk about the importance of place, the necessity of belonging to a place, let us be clear that in Landscape Three place means the people in it, not simply the natural environment. [33]

Landscape without landscape architects is an existing world. A world where people's needs and wishes are translated by hand. A world where contemporary Landscape Architects should experience and learn from to make their own.

2.9 The Value of Vernacular Landscape

The significance of the vernacular landscape is associated with the consequences of human history, traditions, and the heritage of specific culture. The vernacular landscape offers free lessons of unrecorded history of human beings. The dilemma here is our ability to read and understand such landscape.

The argument goes that if one is interested in the landscape itself; what difference does it make to know who occupied, designed, or shaped that landscape? The thing is, as Samuels (1979) stated:

> It represents a certain pattern, style or motif that emerged in the wake of other patterns, styles or motifs. We can trace its aesthetic and institutional origin and be satisfied that it "derived" under the influence of Chinoisere and physiocratic idealism. [34]

Pierce Lewis (1979) wrote, "...If we want to understand ourselves, we would do well to take a searching look at landscapes." The human landscape is an appropriate source of self-knowledge, according to Lewis, because it is "...our unwitting autobiography, reflecting our tastes, our values, our aspirations, even our fears." [35]

J. B. Jackson (1984) suggested that such an approach will enhance our ability to understand the landscape itself and its beauty. He said, "...I suspect that it is by studying the vernacular that we will eventually reach a comprehensive definition of landscape and of landscape beauty." [36]

In his book, William Hoskin (1955) wrote, "The English landscape itself, to those who know how to read it aright, is the richest historical record we possess."[37]

The main emphasis in his most important landscape studies is upon the morphology of locality. As Hoskin (1955) wrote in his essay:

> The view I see, looking at a wide landscape as an historian, is a view composed in the first place of fields, fields of various shape and size, grouped together in certain ways; a view with open roads or deep set half-buried lanes; of churches with or without spires; of grouped villages or perhaps lonely hamlets, or even single farmsteads; of towns of a certain shape and size—in short, everything that has been humanly made, and not sculptured by nature in the first place. [38]

Therefore, by comprehensive study and analysis of the vernacular landscape, one could formulate and trace the intellectual development of a given society's history. This is always the case where vernacular landscape that is cultural-specific has pure characteristics that have been preserved over the centuries.

2.10 Landscape Pattern and Social Structure

Figure 2 Countryside of Tirgue Mures, Romania

According to Jackson (1960), studding rural landscape should be required in training for the design professions. He stated:

> Communities and landscapes have always been organized into patterns, but often by anonymous forces or traditions. The architect or planner who becomes aware of this wide and ancient field of anonymous folk design will learn to see purpose where previously he had seen only disorder; and he will perhaps also see that his own designs are at least in part the expression of his own inheritance. The rural landscape and the rural dwelling are not to be studied as models for imitation, but they reveal how forms and patterns come into being, and the process is by no means always rational. [39]

Therefore, one must always seek "to understand the landscape in living terms, in term of its inhabitant." Judgments about landscape quality must begin by assessing it "as a place for living and working" and proceed toward a conclusion based on how "productive" it is for the need of all of humankind—biological, social, sensual and spirtual.[40]

Jackson insists that one should "experience" the vernacular landscape in order to be able to understand it. To be sure, "experience" here does not mean to have a picnic in the landscape or even to work in the field for hours. Instead, it means what he called for in 1956: "to experience the landscape in term of its inhabitants." He stated:

> Abandoning the spectator stance and…asking ourselves how any man would fare who had to live in it. What chances (for instance) does the landscape offer for making a living? What chances does it offer for freedom of choice of action? What chances for meaningful relationships with other men and with the landscape itself? What chances for individual fulfillment and for social change? [41]

Therefore, landscape evaluation must begin with people, and thus, any definition of "landscape beauty" must incorporate "a new social dimension." This principle of dealing with the landscape "in living terms" has an essential corollary: people must be encouraged to look at their surrounding by themselves and be given the intellectual and physical tools to do much more of the shaping of their own environments. [42]

Furthermore, to understand the landscape in living terms requires crucial attention to the vernacular landscape. For Jackson, the word "vernacular" includes "the environment of workday world." He continues, " The motel, the franchised fast-food stop, and the contemporary house seeking to accommodate new mobile and American recreational lifestyles are as authentic example of what vernacular means as the dwelling of Pueblo Indian or Greek peasant". [43]

In his analysis of what he calls the "Gemeinschaft," or the traditional pre-technological community, the German sociologist Tönnies provides us with a glimpse of the traditional inhabited landscape of Europe four centuries ago:

> The people see themselves surrounded by the inhabited earth. It seems as if, in the beginning of time, the earth itself had brought forth from its womb the human beings who look upon her as their mother. The land supports their tents and houses, and the more durable the houses become, the more men become attached to their own ground, however limited. The relation grows stronger and deeper when the land is cultivated... Even in time of nomadic wandering, family and home are the source of such sentiment... the metaphysical character of the clan; the tribe, the village and town community is, so to speak, wedded to the land in a lasting union. [44]

Jackson (1984) believes that political landscape elements indicate the cultural quality of spatial organization as a "universal need for human nature." He stated that saying "To me, this universal need—and universal ability—to organize space, to divide it into microsapaces, assemble them into macrospaces, is impressive evidence that there is a common, unchanging human nature." Each society in different regions tries to develop their own spatial organization based on their need and the availability of natural resources. [45]

Work cited:

[1] Jackson, J B (1984), *"Discovering the Vernacular Landscape"*, p.7; Yale University Press, New Haven and London.

[2] Taylor, Ken, 'Landscape and Memory, cultural landscapes, intangible values and some thoughts on Asia', pg.1, caneberra act 0200, Australia.

[3] Ibid., p.1

[4] Taun, Yi (1979), 'Thought and Landscape', 89-102 in Meinig D W. 'The eye and the mind's eye' pp. 90 in Meinig ed. (1979) *"The Interpretation of Ordinary Landscapes. Geographical Essays"*, Oxford University Press, New York.

[5] Jackson, J B (1984), p.7

[6] Meinig, D W, 'Reading the landscape' pp. 221 in Meinig ed. (1979) *"The Interpretation of Ordinary Landscapes. Geographical Essays"*, Oxford University Press, New York.

[7] Jackson, J B (1984), p.8

[8] Al-Gilani, Ahmad (1999), '' Creative Landscape design, an Experimental Design process'' p.2

[9] http://conventions.coe.int/Treaty/en/Treaties/Html/176.htm viewed on 20Mar 2011.

[10] Jackson, J B, p.149

[11] http://wordnetweb.princeton.edu/perl/webwn?s=vernacular, viewed on 15 Feb 2011.

[12] http://en.wikipedia.org/wiki/Vernacular_(architecture) , viewed on 15 Feb 2011.

[13] http://www.carterjonas.co.uk/our-services/planning/useful-information/jargon-buster.aspx, viewed on 15 Feb 2011

[14] http://www.louisianavoices.org/edu_glossary.html , viewed on 15 Feb 2011.

[15] King, J "Landscape Architecture without LAs",15 Mrch,2008, found at: http://landscapeandurbanism.blogspot.com/2008/03/landscape-architecture-without-las.html , viewed on 15 Feb 2011.

[16] Ibid, p.1

[17] Eben Saleh, (2001) "Environmental cognition in the vernacular landscape: assessing the aesthetic quality of Al-Alkhalaf village, Southwestern Saudi Arabia", Building and Environment, 36, p.970

[18] P. Groth and T. Bressi, *"Understanding Ordinary Landscapes"*, Yale University Press: New Haven, 1997 (p. 4-5)

[19] Eben Saleh Mohammed A., (2001) *"Environmental cognition in the vernacular landscape: assessing the aesthetic quality of Al-Alkhalaf village, Southwestern Saudi Arabia"*, Building and Environment,36 :965–979

[20] Jackson, J B (1984), p.50

[21] Lewis, P (1979), 'Axioms for Reading the Landscape', 11-32 in Meinig D W. Meinig, D W, 'About the Axioms and about cultural landscape' pp. 12 in Meinig ed. (1979) *"The Interpretation of Ordinary Landscapes. Geographical Essays"*, Oxford University Press, New York.

[22] Quoted by Meinig, D W(1979), p. 198

[23] Lewis, P (1979), p. 14

[24] Meinig, D W (1979), p. 237

[25] Ibid., p.206

[26] Lewis, P (1979), p. 14

[27] Ibid., p.15

[28] Ibid., p. 15-26

[29] Ibid., p.18

[30] Samuels, M, (1979), 'The bioghraphy of Landscape', 51-88 in Meinig D W. Meinig, D W, 'Conclusion' pp. 81 in Meinig ed. (1979) *"The Interpretation of Ordinary Landscapes. Geographical Essays"*, Oxford University Press, New York.

[31] Ibid., p. 81

[32] King, J. "Landscape Architecture without LAs",15 Mrch,2008, found at: http://landscapeandurbanism.blogspot.com/2008/03/landscape-architecture-without-las.html ,viewed on 18 Aug.

[33] Ibid., p. 81

[34] Samuels, M. (1979), 'The bioghraphy of Landscape', 51-88 in Meinig D W. Meinig, D W, 'Conclusion' pp. 52 in Meinig ed. (1979) *The Interpretation of Ordinary Landscapes. Geographical Essays*, Oxford University Press, New York.

[35] King, J.(2008)

[36] Jackson, J. B. (1984), p.149

[37] Meinig, D. W. (1979), p. 195

[38] Quoted by Meinig, D W (1979), p. 205

[39] Ibid., p. 220

[40] Ibid., p. 228

[41] Quoted by Meinig, D W (1979), p. 224

[42] Ibid., p. 224

[43] Ibid., p. 228

[44] Quoted by Jackson, J B (1984) p. 43

[45] Ibid., p. 28

C h a p t e r 3

VERNACULAR LANDSCAPE IN SAUDI ARABIA

Figure 3 Southwestern region of Saudi Arabia, Al-Baha city

3.1 General Background

The vernacular landscape in the southwestern region of Saudi Arabia has a unique character. Due to its location on the western escarpment, this region has different geology, topography, soil, and climate, which create very distinguished landscape characteristics comparing to the rest of the country. This region was an optimum place for early immigrant tribes to settle and start permanent communities thousands of years ago.

The inhabitants of southwestern Saudi Arabia developed a strong social structure based on religious rules, tribal conventions, and tribal norms. Such social structure influenced the overall development of the vernacular landscape, consisting of traditional built forms of settlements, the agricultural landscape, and natural environment. [1]

3.2   Islamic Principles

Most of the Islamic laws and regulations regarding land property are not in contrast with the traditional tribal laws and customs.   While the law ensure the right of individuals and supports the importance of community, some of these rules and regulations were applied even in the pre-Islamic era.

3.2.1 Islamic property law

Islamic Law permits individuals to own property and gives them absolute rights of ownership.   However, Islamic Law has a set of conditions for ownership, which include the need for the owner to be in good usufruct of his possessions, and obliged for their utilization following Islamic investment principles.[2] The Islamic law provides strict regulation and establishes the enforcement of certain principles to prevent individual landowners from harming others when using their property.

The Prophet Mohammed prohibited the exclusive use of land by individuals as private reserves, because such reserves were often used to oppress the local people.   Thus, the right of equal access to, and the people's use of, natural resources became one of the fundamental features of Islamic principles. [3]

3.2.2 Land ownership

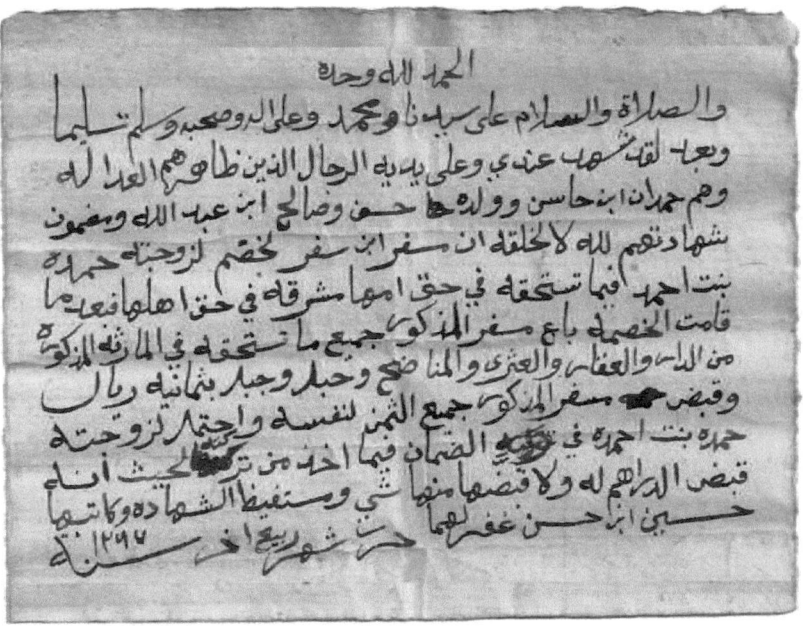

Figure 4 Traditional land ownership document (Hojaj)

According to Lewis (1981), "owning land really meant controlling it, i.e. people only made a claim of ownership to land they could defend." [4] This concept is clearly observed in the southwestern region of Saudi Arabia. Settlements, arable lands boundaries, Hema lands, and defense towers were strategically located to ensure maximum security for the land individuals owned.

According to Ebn Saleh (2002), there are three different kinds of land ownership found in the settlement of the region. The first type is public land, (Hema Al-Qabilah) which belongs to every member of the community. They may gather dry wood, pasture animals, or exploit other natural resources in these areas.

26

Traditionally, such public lands were supervised and managed by an advisory board and the sheik, or tribe ruler.

Other public areas include public open spaces, paths, and public buildings within the built environment. They were mainly supervised and managed by the village board or settlement council. In the traditional tribal society, there were common tribal territories, which had the same characteristics as public land. The boundaries between neighboring tribes were defined by treaties. Within these boundaries, tribe members had a legal right to utilize the land for seasonal agriculture, lumbering, and animal grazing, according to established conventions.

However, these conventions, which defined tribes or villages boundaries, were not permanent due to tribal conflict. The tribal ruler, Al-Sheik, is authorized to grant a piece of public land for any member of the community. After the unification of Saudi Arabia, tribal lands were transformed to public land and under the property of the government.

The second type of ownership is private land. Private land ownership can be established under the reclamation concept. In Islam, anyone who can revive un-owned land for agriculture has the right to develop and own such land. Nowadays, the governmental authorities have restricted this concept to very special cases.

Thirdly, the individual may own land as a result of a personal grant, inheritance, or business transactions between kin group members. Most often, individuals are not allowed to sell land to foreign individuals or investors. Foreign, in this sense, includes people from the next village or non-kin. The inheritance system is a key factor in shaping agricultural land subdivision. [5]

According to Salamon (1993), "the cultural factors shape intergenerational land transmission practices, leading to ethnically distinctive patterns of land tenure, visible in the size of farms, persistence in farming, fragmentation of holdings, and the amount of acreage owned". In the case of inheritance, the landholding could be divided into small tracts with no economic viability, forcing some families to shift production strategies in order to adapt to the new situation and others to resort to migration. [6]

3.2.3 The Hema concept

The word *Hema* (plural *Ahmeyah*) is derived from the Arabic verb *Hama, Yhmee*, meaning to protect.[7] In the southwestern part of Saudi Arabia, people use this term to indicate "preserved land." Hema is a tribal-based land management system defining the use of particular land for specific purposes. Hema might be defined as "a set of regulations controlling the extent and intensity of utilization of resources of a definite place." [8]

Hema is one of the oldest examples of traditional land management systems in history. Most often, Hema is defined in the early stage of the tribal planning process. It was clearly defined for land that has a unique landscape character or special environmental resources. Therefore, Hema lands might be forest, woodland, valley, watershed or even land that distinguishes between two tribes' borders.

As a pre-Islamic principle, the Hema system has preserved the vernacular landscape for thousands of years. It shows how people were dealing with their nature in a sustainable and sensitive manner. There are some surviving Hema in the southwestern region of Saudi Arabia. However, they do not function as Hema and they are no longer owned by the tribe.

3.2.4 Types of Hema

Hema as a concept has been under development for a long time. Location, land use functions, and time periods influence the type of Hema land in history. Therefore, Hema can be classified based on the degree of protection or control and the type of ownership or a social control. Studies about Hema show that the following types existed since earlier times in the Arabian Peninsula:[9]

1. Grazing is prohibited; cutting is permitted during specific periods. This is when plants reach a certain height of growth after they flower and bear fruit. Cut branches are taken outside the Hema to feed the livestock. The tribe council specifies the number of people from each family allowed to do the cutting. Certain trails are specified for the workers in order to prevent the degradation of soil fertility. Certain days are allocated for men, others for women. [10]

2. Grazing and cutting is allowed only after flower and fruits are produced. This allows natural seeding of the soil for the next year or season. Grazing is allowed all year, but the number and types of animals are specified. No restriction on grass cutting.

3. Reserve for bee keeping. Grazing is allowed only after the flowering season. These reserves are closed for five months of the year, including the spring months.

4. Reserve for forest trees, e.g *Juniperus procera, Acacias spp., Haloxlon persicum.* Cutting is only allowed for emergencies or acute needs. [11]

5. Reserve woodland to stop desertification of an area or sand dune encroachment. [12]

Eighmy and Ghanem (1982) added that, Hemas can also be classified by their social control units: tribal, village or individual. A tribal Hema is controlled and used by several villages, which belong to the same tribe. A village Hema is naturally smaller than a tribal one and it is controlled and used by a single village. The individual Hema is usually located next to a cultivated field of an individual owner and is always fenced with stonewalls or a reclaimed land for rain cultivation. [13]

3.3 Natural Transition and Transformation

Evolution is one of the characteristics associated with environmental process. Compared to the built environment, natural landscape has been naturally changed by natural phenomena. Thus, time itself is enough to start the process of the evolution, which consists of many transitional stations.

When humans occupy natural landscape, it becomes cultural landscape by virtue of humans' domination of the landscape and their struggle to control natural processes to meet their needs. Thus, humans and natural forces combine to transform the physical appearance of the landscape.

We should be able to differentiate between natural transition and transformation in the landscape. Long periods of time are needed to change from one state to another and the changes are usually physical. Transformation indicates a dramatic change in physical appearance, while natural transition is more likely a part of a cycle. Natural transition needs substantial periods of time, while transformation requires strong human intervention, and to a lesser extent, time.

One can easily trace the human fingerprint in a cultural landscape, since it leaves physical forms, but it can be quite complicated to determine the human impact on natural processes, which, in a way, influence the contemporary vernacular landscape. In the case of Saudi Arabia, the vernacular landscape in the

30

southwestern region has dramatically changed in the last few decades. Whether it refers to natural or human factors, it is not a part of a natural transition; it is a major transformation in the vernacular landscape.

3.4 The transformation process

The process of change can be observed in human society as well as environment. Some times it is dramatically occurred due to radical changes in political, cultural, economical aspects. Some times it happens gradually that make it hard to be noticed. Thus, Physical change in the vernacular landscape is a consequence of political and social change.

According to Farina (2000), "interactions between society and the environment play an important role in the configuration of "cultural landscapes," i.e. landscapes subject to human influence in which socio-ecological patterns and feedback mechanisms govern biodiversity." [14]

Ebn Saleh (2002) argued that there are two kinds of change observed in the vernacular landscape in Saudi Arabia. First, there are those changes that are provided for in the existing social structure. So long as the roles themselves continue more or less unchanged, these management systems do not affect the structure of the social system itself. They operate within the existing framework, are resolvable in terms of shared values, and offer no challenge to existing institutions.

The second kind of change is in the character of the political system itself; some of its constituent institutions are altered, so that they no longer mesh with other institutions as they once did. This is structural or radical change, and the change to which it gives rise are observable in terms of the existing values of the society. Structural changes engender new types of management, and tradition provides neither precedents nor cures for them. [15]

3.5 History of Transformation

Figure 5 Historical image for Al-Baha city, Saudi Arabia, source (www.ibtesama.com)

There is a critical void in the literature regarding the change in the vernacular landscape of the southwestern region of Saudi Arabia. Oral history and visual assessment could be sufficient tools to trace the physical change in the appearance of the landscape. This could reduce speculation and attribute the change to a sequence of events rather than specific dates.

After the unification of Saudi Arabia in 1932, the tribal system was no longer function as a local government. Instead, national government established new authorities and institutes to implement government plans and policies. It is worth mentioning that some of the tribes in the southwestern region fought against the unification movement. Despite the discovery of oil in 1938, there were no major changes in the people's daily lives and no change can be observed in the vernacular landscape.

One of the most important time shifts to have a dramatic influence on the vernacular landscape occurred in 1957. The government, by royal decree, transformed all tribal Hema to public land. In the event that these lands were forest or arable land, the Ministry of Agriculture and Water would assume responsibility for preserving or developing these lands. Otherwise, it would be the property of the Ministry of Municipalities and Rural Affairs. As a result of this mandate, some of the villages Hema were divided and granted to tribe members to avoid a loss.

In the mid-1970s, political stability and economic growth encouraged people to participate in government agency programs and rely less on traditional work, including agriculture. With population growth and an absence of a master development plan, urban sprawl consumed forests and formerly arable land.

In the 1980s, many of the youngest generation had moved to the main cities to complete their education or to find appropriate jobs. In addition, local markets were full of imported livestock from all over the world. The profitability of agriculture as a business become less certain as the industry became more competitive, especially with the shortage of irrigated water. As a result, large areas of cultivated land were abandoned and most of the cultivated terraces were declined.

This was a critical period in the region's history. The dynamic change in the economic, social structure and the absence of government intervention led to dramatic change in the vernacular landscape.

3.6   Main factors of transformation

3.6.1   Political factors _ landownership

Figure 6 Raghdan Forest (formal Hema transformed to public park)

After the unification of Saudi Arabia in 1932, the government gradually took over the responsibility of managing the vernacular landscape. The change in the landownership system in 1957 has been one of the most significant political interventions since in recent times.   The change dramatically influenced the appearance of the vernacular landscape.

According to Al-Gilani (1998), the government has made this decision for two reasons.   First, it "aimed to give the government a stronger position in natural resource management and reduce tribal conflict."   Secondly, it "aimed to reduce tribal authority, as a part of detribalization policy adopted at that time." [16]

34

As a result, all tribal lands, including Hema, were assigned to the Ministry of Agriculture and Water. Since these lands held substantial forests and arable land, the local municipality sought to own some of this Hema, such as in the case of Raghdan forest, turning it into a public park..

These "public parks" had no management or development plans, which could provide more time for the public to destroy the land over time. Instead of a uniform and properly structure land management process, random construction and tourist activities have badly affected the quality of the forest and its wildlife.

The governmental authorities neither took advantage of this tradition-based land management system nor implemented contemporary environmental planning processes to protect these lands. It is disheartening for the tribes to see their Hema, which were preserved for hundreds of years, destroyed in such a short period of time—in just one generation. Lately, the governmental authorities have recognized the consequences of this decision, and hopefully they can offer more space for public participation in the early stage of the planning process to ensure preserving the rest.

3.6.2 Social factors _ education

The change in the vernacular landscape is usually associated with socioeconomic changes. It is evident that demographic change, economic growth, and changing lifestyles would directly influence any vernacular landscape associated with it.

Saudi Arabia, in the last few decades, has witnessed rapid change in most of its social aspects, which changes the vernacular landscape. Education is one of the most important social aspects and has indeed influenced the vernacular landscape in the southwestern part of Saudi Arabia. Most of the inhabitants would not recognize this aspect as a major factor in the transformation process.

For traditional agriculture-based communities, family members play an important role in the processed of agricultural activities. As a home-based industry, family members are the labor force and have clear tasks ranging from fieldwork to the weekly market. Therefore, any changes in the production line will eventually lead to a deficiency in the family's productivity and could threaten the family business's solvency.

As a part of a strategic educational plan, the central government established new schools for primary education across the country. In the southwestern region, the first official school was opened in a village called "Aldafeer" in 1934.[17] Later, the first educational body was formed in 1955, which supervised 13 primary schools. By 1965, more than 94 schools were opened to offer free education. [18]

Families have not, however, encouraged their children to be enrolled in public education. First, the awareness of the importance of education through government-supported institutes in Saudi Arabia was too low. Secondly, and more importantly, families realized that if their sons went to school, they would have less help working the fields. Over time, however, school enrollments have increased (especially among males), which will have had an influence on the agricultural activities in the region. After graduation, many students work for the government or travel to other cities in pursuit of higher education.

As a result of these factors, agriculture became less profitable, with the exception being for those who had extremely large tracts of productive land and who could afford to hire external (non-family) labor. In this period, most of the traditional practices were lost and cultivated land has been abandoned for the first time. This signals a very critical time in the transformation of the vernacular landscape

3.6.3 Spatial factors _ Road Networks

Figure 7 King Fahad Road, Al-baha, Saudi Arabia

In the southwestern part of Saudi Arabia, people have developed their own network trails based on human and animal scales. The right of way is clearly defined as one of the traditional Islamic principles. For instance, roads should be sufficient for two persons, pulling their camels, passing in different directions.

The traditional trail networks show that people in the mountainous areas are very keen to find the shortest distance, convenient enough to link to their area of interest. As a result, very intensive network patterns, which are completely integrated with the landscape, were established and still in use in some cases.

However, when cars came to the region, the necessity for road expansion became reality. Car roads, with their typical physical features and infrastructure, have had considerable influence on the vernacular landscape in this region, as in most others. First, car road infrastructure has directly influenced the appearance of the landscape as a form of the built environment.

37

Second, in mountainous areas, the construction process requires a lot of excavation and explosive work, which are associated with significant topographical changes in the region. Thus, the water drainage lines, surface runoff, and even the ground water system have been affected by road construction.

It is worth mentioning here that preserving the vernacular landscape was too advanced a principle for the local planning authority to consider at that time. However, the road construction process is one of the hidden factors to have had a major effect on vernacular landscape in the region.

3.6.4 Environmental factors _ climatic change

The southwest region of Saudi Arabia catches the highest amount of precipitation compared with the rest of the country, due to its topography and location within the subtropical zone. The annual precipitation ranges from 120 mm on the coastal side up to 600 mm in the mountains. [19]

These figures vary by elevation and topography. According to Al-Mazroui (1998), the maximum mean annual rainfall in the region occurred at areas called Nimas (512 mm), Biljurashy (430 mm) and Al-bha (324 mm). Also, rainfall occasionally reaches as much as 900 mm in Biljurashy, the area of the case study.[20]

Al-Mazroui's climatological study and rainfall analysis were based on monthly records of rainfall data for 23 years (1970 to1992). The records were obtained from the Hydrology Division of Ministry of Agriculture and Water (MAW) and Meteorological Environment and Protection Agency (MEPA). The data were collected from 30 selected stations distributed across the southwestern region of Saudi Arabia.

The study also indicated that the annual rainfall has substantially decreased in the last few decades. He stated that "it is generally the case, variability increases with decreasing rainfall. On both sides of the escarpment, the variability coefficient is high." [21]

There is no doubt that the amount of rainfall received by an area will change both natural and cultural landscapes. Most of the inhabitants of the region are convinced that agriculture has been abandoned in response to climate change. In fact, climatic change is just one factor in the transformation of the vernacular landscape.

3.6.5 Spatial factors_ building materials

Figure 8 Traditional built environment, Al-Baha city, Saudi Arabia

The built environment is one of the main components of the vernacular landscape setting. The principal is that any change in the construction materials, architecture style, or building regulation can be clearly observed in the vernacular landscape. Thus, in order to trace the change in the vernacular landscape, one should have the basic knowledge about the transformation in the vernacular architecture.

In the southwestern region of Saudi Arabia, the vernacular architecture has a unique character. Over time, buildings have been constructed out of stones, which are the natural stones available locally. The construction process requires cooperative effort from all village members or families. This built environment added to the beauty of the vernacular landscape.

Since houses were built with natural material, aligned to the mountains, and nicely integrated with the natural background, it fits well with the natural landscape. However, this romantic story of the vernacular landscape, including vernacular architecture, has a twist. The introduction of cement as a construction material to the region has dramatically changed architectural styles and the landscape scenery.

According to Al-Haseel (2005), the cement, Egyptian made, reached the region in 1951 along with the brick template. The first building built by concrete blocks was erected in 1952. It was a government building designed for health care. Notably, an agriculture school rented the first complete concrete building in the region.[22]

As a result of the economic and population growth coupled with the lack of building regulation, people built with no respect to traditional architectural styles or formal land use zones. Unfortunately, most of the old buildings were abandoned or have been destroyed to make way for concrete summerhouses.

This new building materials has changed the size, shape, and height of the man-made environment all over the world. There was, however, a chance to use these new materials to preserve or enhance the aesthetic quality of the vernacular landscape and satisfy local needs.

Work cited:

[1] Eben Saleh, Mohammed (2002) 'A Transformation in the Vernacular Landscape of Highlands of Southwestern Saudi Arabia', International Journal of Environmental Studies, 59:1, p.43

[2] J.N coulson, (1964) "A History of Islamic Law", Edinburgh University Press, Edinburgh. (quoted by Ebn Saleh (2002), p. 34)

[3] Eben Saleh, Mohammed (2002), p. 34

[4] Lewis, B., (1988) "The Political Language of Islam", The University of Chicago Press, IL. (Quted by ebn saleh (1995), 82)

5 Eben Saleh. Mohammed(1996) 'Al-Alkhalaf vernacular landscape: the planning and management of land in an insular context, Asir region, southwestern Saudi Arabia', Landscape and Urban Planning 34, p. 82

[6] Salamon, S. (1993), 'Culture and agricultural tenure. Rural Social', 58: 580-598. (quoted by Eben Saleh, (1996), p. 83)

[7] Al-Gilani, Ahmad (1998), "The environment: Theories, Assessment Techniques, and Policies. The Saudi Experience" (unpublished), Ph. D. dissertation, (The University of Edinburgh), p. 170

[8] Eben Saleh, Mohammed (2002), p. 34

[9] Gari, Lutfallah, "Ecology in Muslim Heratige: A Hisory of Hima Conservation System", Environment and History ,The White Horse Press, Cambridge, UK, Vol.12, No.2 (May 2006)found at: http://www.muslimheritage.com/topics/default.cfm?articleid=916#ftn5 viewed at 18 March 2011

[10] Omar, Draz (1976) "*The Hima in the Arabian Peninsula*" , *al-'Arabi* (Kuwait), in Arabic, 211, p. 52-9. (quoted by Gari)

[11] S. Nomanul-Haq (2003), " Islam and Ecology: Toward Retrieval and Reconstruction", in *Islam awl Ecology,* edited by R. C. Foltz *et al.* Cambridge, Massachusetts: Harvard University Press, p. 121-54. (quoted by Gari)

[12] Omar, Draz (1976) op. cit., "the hema", p. 55 (Quoted by Gari)

[13] Eighmy, J.L. and Ghanem,Y.S. (1982), 'The Hema system: prospects for traditional subsistence systems in the Arabian Peninsula'. Culture and Agriculture, 16, p. 10-15

[14] Farina A. (2000), 'The cultural landscape as a model for the integration of ecology and economics', BioScience 50(4):313–320. Found at: http://www.udg.edu/portals/92/Bio%20Animal/pdf/MRD2005.pdf viewed 26 March 2011

[15] Eben Saleh, Mohammed (2002), p. 46

[16] Al-Gilani, Ahmad (1998), p. 181

[17] Ministry of Interior (1989), "*Al-Baha on the Road of Prosperity*", Exclusive summary for King Fahad visit, p. 29 (In Arabic)

[18] Ministry of Culture & Information, Media affair (1988), Al-Baha , "*Summer and Winter place*", p. 37 (In Arabic)

[19] Subyani, Ali (1999), 'Topographic and Seasonal Influences on Precipitation Variability in Southwest Saudi Arabia', JKAU:Earth Sci.,Vol.11, p.90

[20] Mazroui, M (1998), 'Climatological study of the southwestern region of Saudi Arabia. I. Rainfall analysis', Climate Research, Vol.9: 217,219

[21] Ibid., p.219

[22] Al-Haseel, Saed (2005), "*Ghamed and Zahran Urban Treasures*", Al-Madina Publisher Institute, p. 311(In Arabic)

CASE STUDY: HEZNA VILLAGE

Figure 9 Hezna Mountain, Hezna village, Baljurashi

4.1 Introduction

The study area includes one village, Hezna, as a prototype for the transformation of the vernacular landscape in the southwestern region of Saudi Arabia. In the last few decades, the village has undergone dramatic changes in many aspects, which have influenced the appearance of the vernacular landscape. The verification of this assumption was approached through objective analysis and field investigation of the study area focusing on its functional aspect, historical development, and spatial qualities. In addition, oral interviews covered the lack of written materials regarding historical degradation of the agriculture fields. Visual assessment, using photography and updated aerial photograph, can be the physical evidence of the change in the vernacular landscape.

4.2 Location:

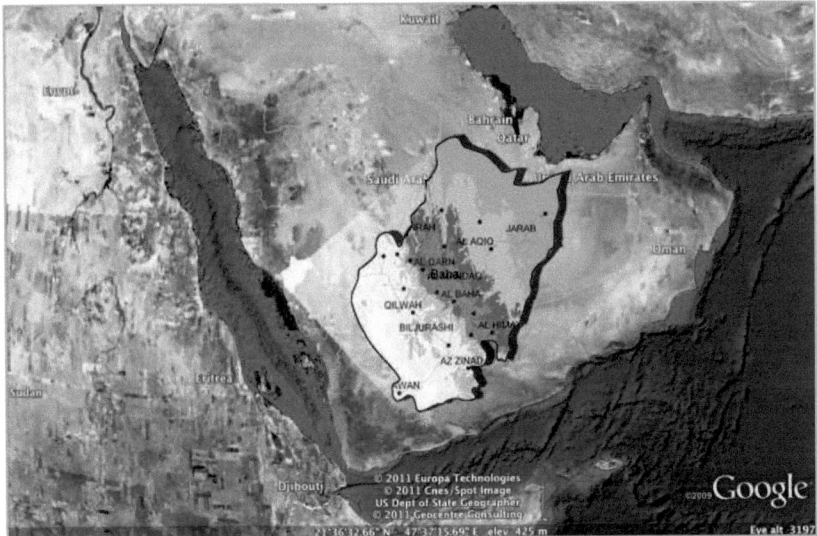

Figure 10 Map of Saudi Arabia (location of the study area)

Hezna village is located in Baljorashi city, the second largest city in the Al-Baha region. Al-Baha is a province located in the southwestern part of Saudi Arabia (see Fig. 10) between longitudes 41/42 E and latitudes 19/20 N. It has an area of 15,000 km², and a population of 459,200. [1]

4.3   Topography

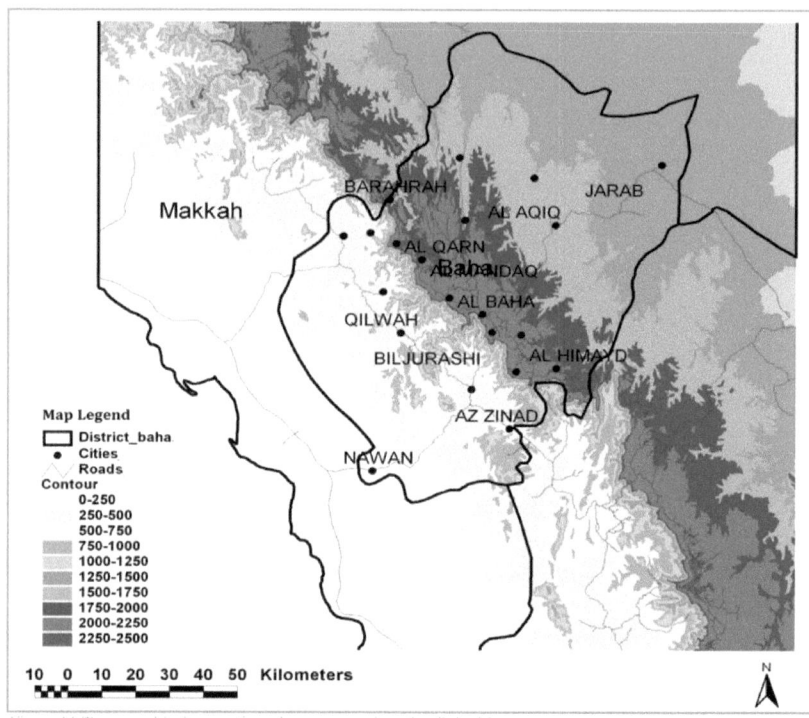

Figure 11 Topographical map of southwestern region, Saudi Arabia

The Al-Baha region is divided by large rocky steeps into two main sectors. To the west is a coastal plain, Tihama. It is a lowland area, characterized by very hot and humid weather with very little rainfall. To the east is the mountain range of Al-Sarat (where the case study is located), with an elevation of 1,500 to 2,450 m above sea level. It is characterized by high mountains with temperate weather and rich plant cover resulting from a relatively high annual rainfall. [2] (see fig. 11)

Hezna is located at the edge of the Sarat escarpment and has the highest altitude on Hezna Mountain. Its location at edge of the Sarat escarpment provides sufficient cultivated land in the valleys and in the surrounding terraced mountains.

45

4.4 Climate

The variation in topography influences the climate of the region. The al-Sarat area is exposed to the formation of clouds and fog, which often occurs in winter when air masses arrive from the Red Sea, accompanied by thunderstorms. In spring and summer the climate is mild and pleasant. The climate in the Tihama area is different from that in al-Sarat, although they are separated by no more than 25 km. Tihama is a high-variability coastal plain: hot in summer, warm in spring and mild in winter. The Scarp Mountains, which are characterized by high altitude, tend to have a lower annual range of temperature than the surrounding low areas. The mean minimum temperature is as low as 0°C in scattered locations, especially in high peaks in winter. The mean monthly maximum temperature is 25°C in summer. The relative humidity ranges from 35% in summer to 65% in winter. [3]

4.5 Traditional local government and social structure

The tribal organization in the highlands of southwestern Saudi Arabia is governed by a hierarchy of political power. Every main tribe is composed of sub-tribes, the sub-tribes are composed of clans, and clans are composed of luhmah (kin-groups). [4]

In the village scale, which belongs to one tribe, the village ruler (Al-Arifah) is the one responsible for managing daily business, including disputes within his village boundaries. In terms of foreign affairs (dealing with a neighboring village, for instance), Al-Arifah and the village council are the main representatives in the tribe assembly. The village council consists of different (Naib), elected representatives of every kin-group in the village. Every kin-group consists of different families who are involved in agriculture and settle in a village of defined boundaries and properties.

In the case of Hezna village, five kin-groups still exist in the village. Traditionally, the management of agricultural activities and water management start from the kin-group and family level. Usually, every family in the kin-group tries to acquire land beside each other to ensure the maximum advantages of the extended land. The influence of the kin-group basically depends on how much cultivated land they own. The more land they own, the more power and influence they have on local affairs.

This type of social structure is one of the key factors in the preservation of agricultural practices in the region. It is very significant to clarify this political hierarchy in order to understand the development of agriculture pattern and to trace the transformation of the vernacular landscape in the region.

4.6 Sustainable Traditional Practices

Over time, inhabitants of southwestern region of Saudi Arabia developed a cumulative knowledge. It is reflected in their land use pattern and agricultural practices, which are considered a remarkable example of sustainability and environmental sensitivity. The relation between humans and nature has reached a mature level where people do not live from nature, but they live for nature.

It is quite logical that people appreciate nature since it provides them with basic human sustenance. But what is remarkable here is the degree of understanding of the environmental process without any formal educational background. They know exactly when and how they should react to help the ecosystem to recover by itself.

The principle of preserved land "Hema," with its rules and fines, provides an excellent example of a sustainable land management system. Just because they have some memories and childhood stories, People could modify some building to save an old tree while the next generation of the same people cut all of the trees down to build a summerhouse. They used to pray for God to save their crops from damage or flooding. They use to leave good deal of their crops for birds and wild animal, how could they kill them for nothing.

In terms of water management, the traditional practices indicate a high degree of efficiency and cooperative spirit among community members. In the rain season, they build hundreds of small dams in their farmlands to ensure the maximum replenishment of underground water. Also, they have the same system for well water resources. There is a man-made ditch between wells, especially in agricultural fields, to divert extra water to the neighboring well. In addition, there were designed places to collect the water for domestic and wild animals. In the rain season, people collect the rainwater from the roof of the building, not as drainage, but for personal use.

Most of the traditional practices were lost due to the use of machinery and the abandonment of agricultural activity. Unfortunately, new generations never experience such practices; those alive today must simply enjoy these tales as heritage stories from our elderly kinfolk.

4.7 Visual & spatial quality

The spatial organization in this region has been developed in respond to social needs, physical landscape and the capacity of the natural recourses. As a result, the main land use pattern can be typical for most of the villages in this region.

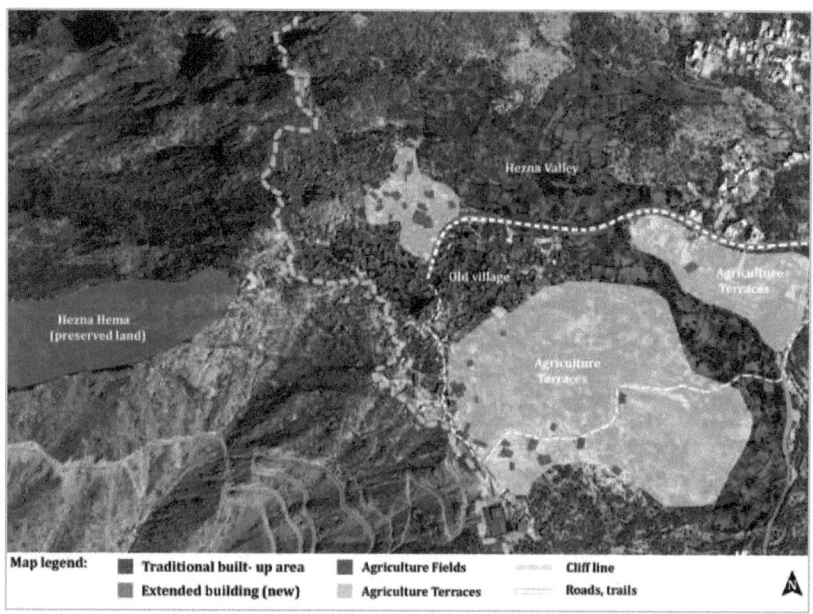

Figure 12 Land use map of Hezna village

The land use pattern of Hezna village can be classified as follows:

1. Natural Landscape

2. Built Environment

3. Agricultural Fields

49

4.7.1 Natural landscape

Figure 13 Natural landscapes, Hezna village

- **Natural forest**: Usually preserved by local people as Hema. Natural forests are rich in trees like Juniper, Acacia and wild olive trees. If no forest exists, Hema could include pasturelands, arable lands or watershed. Hema boundaries are clearly defied and well known by local people.

- **Pasturelands**: Located in the ridge land around the village. In some cases, like Hezna village, they use the Hema for grazing after seasonal rain.

4.7.2 Built environment

Figure 14 Built environment, Hezna village

In the early planning stage, the build-up area is strategically located in the village. Local inhabitants build their houses in the highlands of the mountains to gain better visual access to the agricultural fields and the village boundaries. It is the optimal place to avoid heavy rain, flood, and external attacks. Also, it is close to a main source of construction materials. The inhabitants defined a buffer zone between the built-up area and the agricultural fields to manage further extension of the residential zone. The mosque is the center of the built environment and the main meeting point for the villagers. The vernacular architecture here is completely integrated in the landscape with high aesthetic quality.

4.7.3 Agricultural fields

Figure 15 Agricultural fields, Hezna valley

The agricultural activity in the area is managed by a very interesting and complicated set of rules and customs. The agricultural fields can be classified by their size, location, irrigation system, and the type of ownership. In general, the agricultural terraces and the agricultural fields are the main categories for private agriculture.

- **Agriculture terraces (Alshea'ab):** Includes all types of farmland that is rain fed only (Athery). They are sloping terraces managed by one or more families. In this case, man-made ditches were built, parallel to the contour lines, to collect the rain runoff water as the main source of irrigation. In these terraces, wheat and corn are the main crops. Agricultural terraces are the first farmlands to be abandoned or destroyed due to the shortage of rainfall, erosion, and the amount of effort needed to maintain them.

- **Agricultural fields (Albilad):** Includes all types of farmland that is irrigated mainly from wells (masqawi) and rain runoff as a second source. They usually share water for more than one plot based on a priority schedule. These fields are located in the main valley of the village. They comprise a larger area compared to the slope terraces.

Local people distinguish their farmland by its size and location with different names. (Al-qasabah) is the smallest cultivated unit (2mx2m). It is not the lowest land in the valley and often it has natural trees around it. Then, there is (Al-sheti) farmland (2mx 10m). This can have different shapes and locations in the valley. The biggest farmland unit (5mx10m) is called (Al-falaj) and is located close to the water source.   For local people, the term (Al-bilad) indicated all types of cultivated land. Since it has the best shape and size, it is the optimum place for vegetables and fruits trees to be cultivated. Some fruit trees grow naturally here, such as pomegranate, peach, and fig trees. Unfortunately, most of the agricultural fields were abandoned. However, fields near the village are last to be abandoned and some are still cultivated today.

4.8 Assessment of vernacular landscape

4.8.1 Historical degradation of agriculture fields in Hezna valley

The historical degradation of agricultural fields is considered one of the most important factors in the transformation process of the vernacular landscape. Field investigation and oral interviews with elderly farmers has been conducted in order to verify the transformation history. In Hezna valley, the transformation process started around 1975. (See fig. 16)

| Map legend: | Cultivated land | Abandoned 1995-2010 | Abandoned 1985-1995 | Abandoned 1975-1985 |

Figure 16 Degradation map of agricultural fields, Hezna valley

From 1975-1985, most of the traditional agricultural fields were abandoned, for the first time, due to the death of a landlord or shortage of water. This is very critical time shift in the transformation history and it's hard to refer it to one aspect. However, This came after 20 years of the establishment of 94 schools in the region.

From 1985-1995, some lands were still cultivated. However, especially in this period, most of the traditional agricultural practices were lost. Farmers began using machinery and chemical fertilizer.

From 1995-2010, the agricultural activity had a completely different intensity and function. Most of the farmers, or landlords, had different income from jobs or business activities. Therefore, the agricultural fields, in this period, functioned as large kitchen gardens for private use not connected to the weekly market.

Nowadays, only 4-5 plots are still cultivated for private use in Hezna valley. Three of the farmers are old farmers who have "nothing to do in the morning except for practicing farming." In some cases, an external labor force was brought in to cultivate the land. The role of agriculture in the village community has changed from production to pleasure, from a survival-oriented occupation to fitness training.

4.8.2 Comparative analysis of vernacular landscape components:

This analysis is an attempt to compare the traditional and contemporary vernacular landscape. From traditional and existing conditions, a simple evaluation examines the influence of different components on the transformation of the vernacular landscape. The main components include natural and cultural landscape, the built environment, and cultural significance. Every element of these practices has been evaluated in terms sustainability, environmental friendly practice, aesthetic quality, and visual impact on the vernacular landscape.

Vernacular landscape components	Traditional practices				Contemporary practices			
Evaluation criteria	S.	E.F	A.Q.	V.I.	S.	E.F	A.Q.	V.I.
Natural, cultural landscape								
• Natural vegetation								
• Agriculture fields								
• Water management								
• Wild life								
Built Environment								
• Construction materials								
• Architecture style								
• Location suitability								
• Buildings character								
Cultural significant								
• Political hierarchy								
• Landownership system								
• Economic growth								
• Social & spiritual value								

Table legend:

S.	Sustainability		Positive impact
E.F.	Environment Friendly		Negative impact
A.Q.	Aesthetic quality		No impact - NA
V.I.	Visual impact		

56

4.8.3 Visual assessment:

This visual assessment shows the panoramic view of Hezna valley towards the village. (See fig.17) The panoramic view has been selected because it shows the traditional agricultural fields, built environment, and the natural landscape. It is not the aim here to assess the aesthetic quality of the landscape. Instead, the main aim is to evaluate simply which elements have a negative or positive impact on the appearance of the vernacular landscape of the village.

Figure 17 Panoramic view of Hezna valley.

Positive visual impact elements:
    1. Traditional defense tower (stone structure)
    2. Natural vegetation (Juniper, Acacia trees)
    3. Cultivated agriculture terraces
    4. Random building, trees (farmer shelter)
    5. Natural rocks hill
    6. Traditional stone structure
    7. Natural vegetation (Hezna Mountain)

Negative visual impact elements:
    1. Abandoned agricultural fields
    2. Summerhouse (traditional arable land)
    3. 7 floors of a private building (highest building)
    4. Telecommunication tower
    5. Water tank
    6. 4 floors of a private building (on traditional agricultural fields)

4.8.4 Future of the vernacular landscape

The historical degradation map shows a constant decline of agricultural fields. As a result, the transformation of the vernacular landscape in Hezna is highly influenced by the physical change of the agricultural fields in the valley. On the other hand, a new generation of natural vegetation has grown in the agricultural fields and terraces in the last few decades. In addition, new buildings have grown to reach 7 floors on the traditional arable land.

The change in agricultural activities is considered the main factor influencing the appearance of vernacular landscape of Hezna. Therefore, it is significant to illustrate how this valley would look 50 years ago. This could help to imagine how it will look in the next 50 years. The traditional scenario illustration was reviewed by an elderly farmer in the valley. The future scenario assumes that no major changes will occur in the political, economic, and social levels.

Figure 18 Illustrated aerial image of Hezna valley 1960

Figure 19 Illustrated panoramic view of Hezna valley 1960

Figure 20   Aerial image of Hezna valley 2011

Figure 21 Panoramic view of Hezna valley 2011

Figure 22   Illustrated Aerial image of Hezna valley 2060

Figure 23 Illustrated panoramic view of Hezna valley 2060

59

4.9 Recommendations and Remarks

The transformation process of the vernacular landscape has taken over 30 years. Thus, there are no feasible quick solutions to change the whole situation in the near future. Instead, long-term planning and vision is needed to restore and preserve the vernacular landscape.

The transformation in Hezna could be a prototype of the transformation in the region. However, there are some villages in which the transformation of the vernacular landscape is worse or somewhat better. Essentially, the reasons behind the transformation are the same. The influence of different factors could vary from one village to the next. For instance, the transformation of Hema land to public land in Hezna village has no great influence in the overall vernacular landscape compared to the degradation of agricultural fields due to socioeconomic factors. However, Hema transformation is the main factor of transformation in some villages.

As mention earlier, the main factors of the transformation are related to political, economical, social aspects. Therefore, mulit- disciplinary effort should be made to have a comprehensive proposal coming out from a holistic approach.

From landscape architecture's point of view, recommendations and polices can be developed for the natural landscape, built environment, and agricultural fields to enhance and preserve the overall vernacular landscape. Some of these might include the following suggestions.

In terms of the natural landscape:

- Preserve the natural landscape through the revival of the Hema concept and integrate it with formal planning policies. In this region, there is no chance to preserve land unless local people are involved and are determined to protect it.

- Protect wildlife in the natural landscape, define endangered species, connect natural forest for wildlife corridor, regulate hunting activities, and encourage traditional practices that enhance wildlife diversity.

In terms of the built environment:

- Establish detailed master plan, define appropriate built-up area, create alternative zone for urban sprawl, preserve traditional arable land and agricultural fields, and define vernacular and rural building character; specifying architectural styles, height, shape, etc.

- Encourage green architectural principles, using local materials for construction, environmental friendly designs, and adapt traditional building practices.

In terms of agriculture:

- Establish new "agriculture committee" with a rescue mission that involves local stakeholders and all related governmental authorities, such as the Ministry of Agriculture and Ministry of Water, and local municipalities.

- Form non-governmental organizations (NGO) responsible for agricultural activities and authentic representative of local farmers.

- Comprehensive subsidies program for agricultural activity, encourage local and bio products, encourage traditional agriculture practices, and resolving the landownership conflict.

- Ensure a high degree of public participation in the early planning stage regarding the management of agricultural activities.

- Prepare an appropriate investment environment for companies interested in agriculture and develop a new regulation for buying the right for land development.

- Encourage eco-tourism and agro-tourism activities in the region.

- Establish an advanced awareness program using media and tribe assembly to show the importance of agriculture and the consequences for abandonees.

- Encourage working groups and scientific research in collaboration with universities and interested research centers.

4.10 Conclusion

The transformation of vernacular landscape has drawn the attention of many as a worldwide phenomenon. The exceptionality in the case of Saudi Arabia lies in the country's rapid developments due to economic growth, political stability, and population growth. As a result, the transformation in vernacular landscape occurred dramatically in quite a short time.

The reasons for the transformation vary from village to village and from case to case. In general, the transformation of the agricultural fields is one of the most significant factors influencing the vernacular landscape. Therefore, there is a need to define a new role for agriculture in highly developed traditional society. In the case of Saudi Arabia, this critical issue requires high-level decisions in order for the change to occur.

In the past, administrative agencies were not up to the task of taking advantage of the available planning tools or tribal-based land management systems into contemporary planning processes. Governmental plans were not be implemented effectively because authorities did not involve locals in the decision-making process. As a result, local inhabitants had little interest in participating and their sense of responsibility was lost.

Preserving vernacular landscape is not a call to live in the past or to avoid using new technology or materials. This is a call to take advantage of new technology and contemporary planning tools to preserve our cultural heritage. It is a call, especially for professionals, to dig deep into our traditional environmental practices in order to adapt proposals suitable for this region while satisfying the local people's needs and ambitions.

There is no doubt that vernacular landscape deserves more in-depth study and analysis. Hopefully, this and other studies pertaining to the dynamics behind the vernacular landscape will help to call attention to vernacular landscape that reflects the natural development of the tradition, cultural and history of human being.

Work cited:

[1] http://en.wikipedia.org/wiki/Al_Bahah_Province, viewed on 20 April 2011

[2] Abou Zied, Ehab (2010) 'Effect of the climate and some different protein diets on the visitation pattern of flesh and blow flies of Gebel Al- Baher, Al- Baha Province, Kingdom Saudi Arabia', Egypt. Acad. J. biolog. Sci., 3 (1): 133 – 144, p. 45

[3] Subyani, Ali (2000) 'Topographic and Seasonal Influences on Precipitation Variability in Southwest Saudi Arabia' , JKAU: Earth Sci., vol. 11, pp. 89-102, p.95

[4] Eben Saleh, Mohammed (2002) 'A Transformation in the Vernacular Landscape of Highlands of Southwestern Saudi Arabia', International Journal of Environmental Studies, 59:1, p.45

GLOSSARY

Arifah. Village ruler.

Albilad. Farm lands.

Alshea'ab. Agriculture terraces.

Athery. Local term for rain fed terraces.

Hema. Preserved land, traditional –based land management system.

Hezna. A village in the southwestern part of Saudi Arabia, study area.

Luhma. Kin-group.

Masqawi. Local term for irrigated farmlands.

Naib. Elected representative for villager board.

BIBLIOGRAPHY

Literature:

Abou Zied, Ehab, "Effect of the climate and some different protein diets on the
   visitation pattern of flesh and blow flies of Gebel Al- Baher, Al- Baha Province,
   Kingdom Saudi Arabia", Egypt. Acad. J. biolog. Sci., 3 (1): 133 – 144, 2010.

Al-Gilani, Ahmad, "The environment: Theories, Assessment Techniques, and Policies.
   The Saudi Experience" (unpublished), Ph. D. dissertation, The University of
   Edinburgh,1998

Al-Gilani, Ahmad, " Creative Landscape design, an Experimental Design process" 1999.

Al-Haseel, Saed, "*Ghamed and Zahran Urban Treasures*", Al-Madina Publisher Institute,
   2005. (In Arabic)

Eben Saleh, "Environmental cognition in the vernacular landscape: assessing the
   aesthetic quality of Al-Alkhalaf village, Southwestern Saudi Arabia", Building and
   Environment, 2001.

Eben Saleh, Mohammed, 'A Transformation in the Vernacular Landscape of Highlands
   of   Southwestern Saudi Arabia', International Journal of Environmental Studies,
   59:1, 2002.

Eben Saleh. Mohammed(1996) 'Al-Alkhalaf vernacular landscape: the planning and
   management of land in an insular context, Asir region, southwestern Saudi Arabia',
   Landscape and Urban Planning 34, 1996

Eighmy, J.L. and Ghanem,Y.S., 'The Hema system: prospects for traditional subsistence
   systems in the Arabian Peninsula'. Culture and Agriculture, 16, p. 10-15, 1982.

Farina A., 'The cultural landscape as a model for the integration of ecology and
   economics', BioScience 50(4): 313–320. 2000.

Gari, Lutfallah, "Ecology in Muslim Heratige: A Hisory of Hima Conservation System",
   Environment and History ,The White Horse Press, Cambridge, UK, Vol.12, No.2
   2006.

Jackson, J B "*Discovering the Vernacular Landscape*", Yale University Press, New Haven and
   London, 1984.

J.N Coulson, "*A History of Islamic Law*", Edinburgh University Press, Edinburgh, 1964

Lewis, B., "*The Political Language of Islam*", The University of Chicago Press, IL, 1988

Lewis, P 'Axioms for Reading the Landscape', 11-32 in Meinig D W., 'About the Axioms and about cultural landscape' in Meinig ed. *"The Interpretation of Ordinary Landscapes. Geographical Essays"*, Oxford University Press, New York. 1979.

Mazroui, M, 'Climatological study of the southwestern region of Saudi Arabia. I. Rainfall analysis', Climate Research, Vol.9: 217,219, 1998

Meinig, D W, 'Reading the landscape' in Meinig ed. *"The Interpretation of Ordinary Landscapes. Geographical Essays"*, Oxford University Press, New York. 1979.

Ministry of Interior, "Al-Baha on the Road of Prosperity", Exclusive summary for King Fahad visit,1989. (In Arabic)

Ministry of Culture & Information, Media affair, Al-Baha , "summer and winter place", 1988. (In Arabic)

Omar, Draz *"The Hima in the Arabian Peninsula"* , al-'Arabi (Kuwait), in Arabic, 211, 1976.

P. Groth and T. Bressi, *"Understanding Ordinary Landscapes"*, Yale University Press: New Haven, 1997

S. Nomanul-Haq , " Islam and Ecology: Toward Retrieval and Reconstruction", in Islam law Ecology, edited by R. C. Foltz et al. Cambridge, Massachusetts: Harvard University Press, 2003.

Taylor, Ken, 'Landscape and Memory, cultural landscapes, intangible values and some thoughts on Asia', Canberra act 0200, Australia.

Taun, Yi, ' Thought and Landscape', 89-102 in Meinig D W ed. *"The Interpretation of Ordinary Landscapes. Geographical Essays"*, Oxford University Press, New York. 1979.

Website Links:

King, J "Landscape Architecture without LAs", 15 Mrch,2008, found: http://landscapeandurbanism.blogspot.com/2008/03/landscape-architecture-without-las.html , viewed on 15 Feb 2011.
http://conventions.coe.int/Treaty/en/Treaties/Html/176.htm viewed on 20Mar 2011.
http://wordnetweb.princeton.edu/perl/webwn?s=vernacular, viewed on 15 Feb 2011.
http://en.wikipedia.org/wiki/Vernacular_(architecture) , viewed on 15 Feb 2011.
http://www.carterjonas.co.uk/our-services/planning/useful-information/jargon-buster.aspx viewed on 15 Feb 2011
http://www.louisianavoices.org/edu_glossary.html , viewed on 15 Feb 2011.
http://en.wikipedia.org/wiki/Al_Bahah_Province, viewed on 20 April 2011

Printed by Books on Demand GmbH, Norderstedt / Germany